Popcorn Science

Natalie Lunis and Nancy White

Newbridge Educational Publishing,
a Haights Cross Communications Company, New York

Table of Contents

Popcorn Science
ISBN: 1-58273-371-6

Publisher: Vita Jiménez
Executive Editor: Janet Reed
Creative Director: Ruth Schultz
Senior Production Manager: Anne Brown

Written by Natalie Lunis and Nancy White
Edited by Joanne Allgor
Designed by Frances Schlesinger
Production Manager: Michael Miller

Newbridge Educational Publishing
11 East 26th Street, New York, NY 10010
www.newbridgeonline.com

Photo Credits:
Cover: Larry Lefever/Grant Heilman Photography; Page 1: Elizabeth Hathon/The Stock Market; Page 3: Art Phaneuf/New England Stock Photo; Pa D. Young-Wolff MR/PhotoEdit; Page 5: Runk/Schoenberger/Grant Heilman Photography; Page 6: Jose Galvez/PhotoEdit; Page 7: (ears of corn) Runk/Schoenberger/Grant Heilman Photography, (uses of corn from top) Barry L. Runk/Grant Heilman Photography, Larry Lefever/Grant Heilma Photography, Felicia Martinez/PhotoEdit, Larry Lefever/Grant Heilman Photography; Page 8: Larry Lefever/Grant Heilman Photography; Page 9: Runk/Schoenberger/Grant Heilman Photography, (inset) Arthur C. Smith/Grant Heilman Photography; Page 10: Grant Heilman/Index Stock; Page (clockwise from top left) Barry L. Runk/Grant Heilman Photography, Runk/Schoenberger/Grant Heilman Photography, Runk/Schoenberger/Gran Heilman Photography, Runk/Schoenberger/Grant Heilman Photography; Pages 12–13: Larry Lefever/Index Stock, (top inset) Tom Pantages, (bot inset) Randy Duchaine/The Stock Market; Page 14: Michael Mancuso, Omni-Photo Communications; Page 15: (from top) Thomas Hovland/Grant Heilman Photography, Mug Shots/The Stock Market; Page 16: Elizabeth Hathon/The Stock Market; Page 17: (clockwise from top right) David Lawrence/The Stock Market, Michael Newman/ PhotoEdit, Superstock, Eric Kamp/Index Stock, Wallace Garrison/Index Stock, Scott Campbell/ International Stock; Page 18: Frances M. Roberts

10 9 8 7 6 5 4

Mmm . . . **popcorn**. You have probably eaten it, and you may have even made it yourself. But if you stop to think about popcorn, you are sure to come up with many questions. Are you ready for "popcorn science"?

What Is Popcorn?

You have probably made popcorn or seen it at the store. But where was the popcorn before it was in a jar or package? What is popcorn?

Each fluffy piece of popcorn started as one kernel on an **ear** of corn. Each **kernel** is a corn seed.

ompare the unpopped ernels of corn with the opped kernels. Tell three ays in which they are fferent.

Can All Corn Be Popped?

There are several different kinds of corn, and they all have different uses. Think about the kind of corn that you eat at a cookout. What happens when you heat the corn on the grill? Do the kernels pop?

This chart shows four different kinds of corn and their uses.

Type of Corn	How It Is Used
sweet corn 	cooked for people to eat with meals 
popcorn 	popped for people to eat as a snack 
flour corn 	ground into flour for making cereal, bread, and tortillas 
field corn 	fed to cows, pigs, and other animals 

Corn is very important to people around the world. Use the information in the chart to draw a conclusion about why it is important.

Where Does Popcorn Come From?

Like all corn, popcorn grows on a farm. Corn plants can grow to be up to ten feet tall. Have you ever stood in a cornfield? Were the plants taller than you, or were they not yet fully grown?

Here is a diagram of a corn plant and an ear of corn.

Which part brings water from the soil to the plant? Which part holds the plant up? Which part do we eat?

How Does Popcorn Grow?

Some popcorn kernels are used as seeds to grow more popcorn plants. Farmers use **seed drills** to plant the seeds. This machine digs holes, drops seeds into them, and covers the seeds with soil. It takes about four months for the seeds to develop into fully grown popcorn plants.

These photos show four stages in the growth of a popcorn plant. Describe the changes that you see.

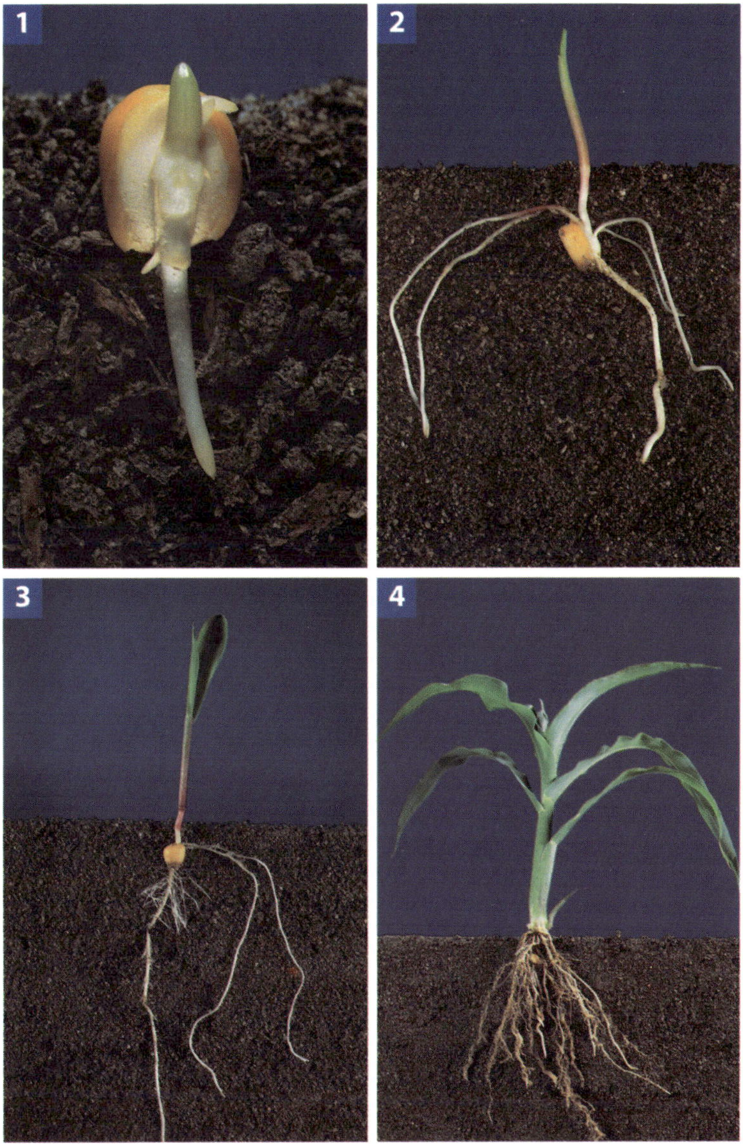

What Makes Popcorn Pop?

Look at the photos. What makes the hard yellow popcorn kernels change into the fluffy white popcorn that you eat?

Each kernel has a tiny bit of water inside. When the kernel is heated, the water turns into steam. The steam builds up until the kernel finally explodes. The explosion blows the kernel inside out.

Based on what you know about unpopped and popped kernels, which one would you guess weighs more?

How Popular Is Popcorn?

Popcorn is one of the most popular snack foods in the United States. Americans use more than 500,000 pounds of popcorn a year!

About three-tenths of the popcorn is eaten at movies, baseball games, and other fun events.

About one-tenth of the popcorn is used as seeds to grow new popcorn plants.

\bout six-tenths of the popcorn is eaten at home.

'his **pie chart** shows the nformation you just ead in another way. Can ou match each section f the chart with the ext? What would you ut in the blank section?

Is Popcorn Good for You?

Almost everyone likes popcorn, but is it a healthy snack or just **junk food**? To answer this question, consider these facts:

- Popcorn is a kind of corn.
- Corn is a **grain**.
- Grains are good for you.

What conclusion can you draw?

Suppose you wanted a nutritious snack. Which of these foods would you choose? Classify the foods in the photos as healthy snacks or junk food.

What did you learn about popcorn science? What surprised you the most? Keep looking around and asking questions. You never know where science discoveries will pop up!

Glossary

ear (EER): The part of a corn plant to which the kernels are attached.

field corn (FEELD korn): A type of corn that is fed to animals.

flour corn (FLOW-er korn): A type of corn that is used to make cereal, bread, tortillas, and other foods.

grain (GRAYN): A plant whose seeds we eat. Corn, wheat, rice, and oats are all grains.

junk food (JUNK food): Food that is not good for our bodies.

kernel (KER-nul): The seed of a corn plant; the part of the corn plant we eat.

pie chart (PYE chart): A round chart that compares different amounts in sections that look like slices of pie.

popcorn (POP-korn): A type of corn that is used for popping.

root (ROOT): The part of a plant that brings water from the soil to the stem.

seed drill (SEED drill): A farm machine used for planting seeds.

stalk (STAWK): The stem of a plant; the part that holds up the plant and brings water from the roots to the rest of the plant.

sweet corn (SWEET korn): A type of corn that is eaten by people.

Index